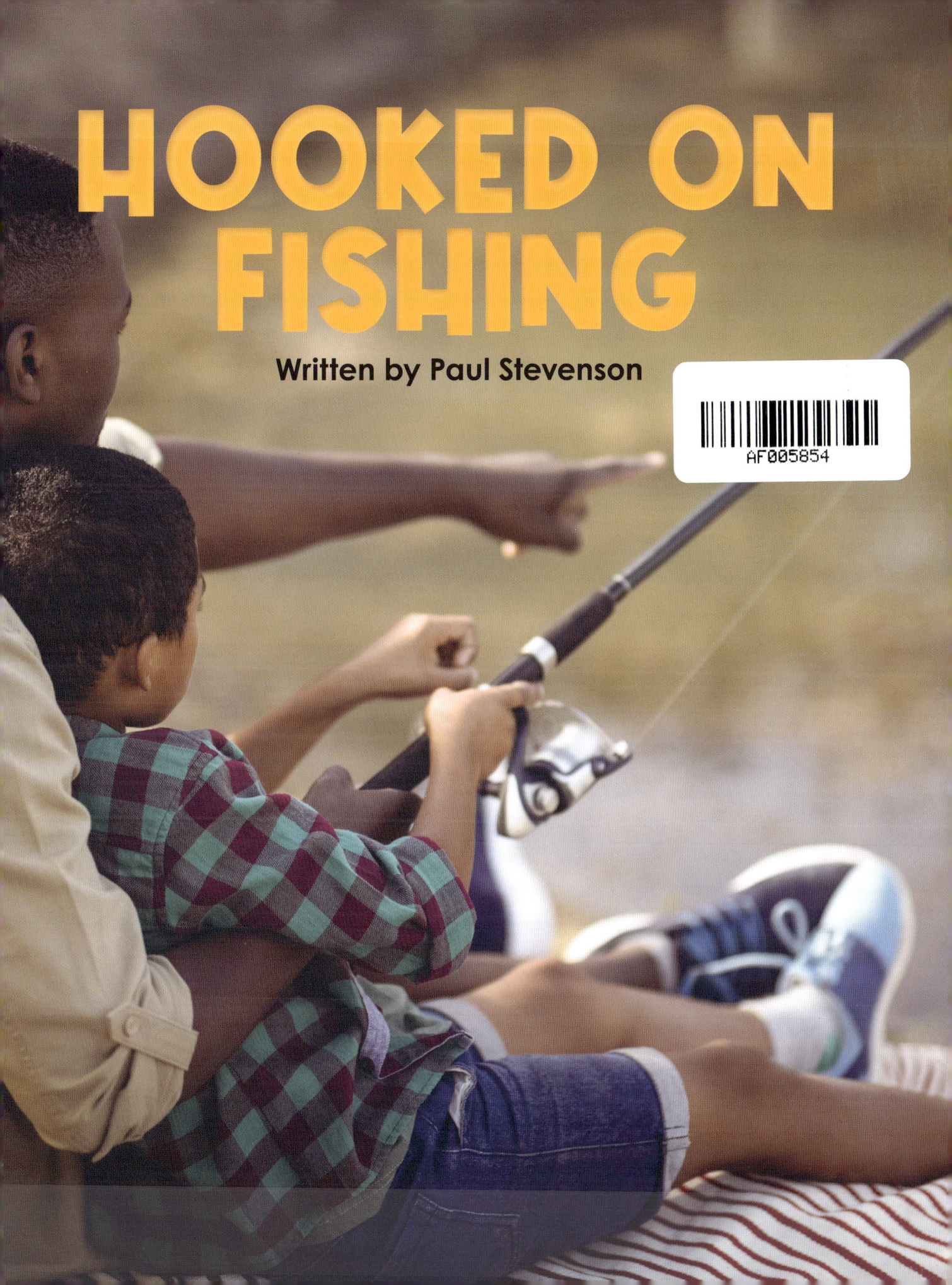

HOOKED ON FISHING

Written by Paul Stevenson

CONTENTS

All words in **BOLD** can be found in the glossary.

Gone Fishing	4
Spin Fishing	6
A Day at the Lake	8
Floats and Sinkers	10
Bait and Lures	12
Caring for the Catch	14
Fly Fishing	16
Fly Fishing – Casting	18
Flies and Gear	20
Sea Fishing	22
Catching a Marlin	24
Ice Fishing	26
Competitive Fishing	28
Catching a Record	30
Glossary	31
Index	32

First published in 2024 by
Hungry Tomato Ltd
F15, Old Bakery Studios,
Blewetts Wharf, Malpas Road,
Truro, Cornwall,
TR1 1QH, UK.

Thanks to our editor, Julie Tofflemire.

Copyright © 2024 Hungry Tomato Ltd

No part of this publication may be reproduced, stored in a retrieval system, or transmitted in any form or by any means, electronic, mechanical, photocopying, recording, or otherwise, without prior written permission of the copyright owner.

A CIP catalogue record for this book is available from the British Library.

ISBN 9781835691205
Printed in China

Discover more at
www.hungrytomato.com

Neither the publisher nor the author shall be liable for any bodily harm or damage to property whatsoever that may caused as a result of conducting any of the activites in this book.

DISCLAIMER:

Always make sure you have the right permissions to fish at a certain spot. Different areas have different rules, and you may need a fishing license. There may also be restrictions on how many fish you can catch and what time of year you can fish.

The techniques in this book have been performed by experienced anglers. Do not attempt them without guidance.

GONE FISHING

Fishing is more than just catching fish. It's the anticipation of the catch. It's watching wildlife while you wait. It's enjoying the outdoors and the fresh air.

Spin fishing is using a rod with a reel that spins around.

Reel

Fly fishing is catching fish with flies, **imitation baits** that look like insects.

Flies

Sea fishing is catching fish from a beach, rocks or a boat.

Mahi-mahi (Dolphinfish)

Fishing is very popular! There are around 4 million **anglers** in the UK.

SPIN FISHING

Spin fishing is a great technique for beginners. It got its name from the spinning reel that is used.

When spin fishing, a lure is attached to the end of the line. The line is **cast** into the water. The angler then starts **reeling in** the lure.

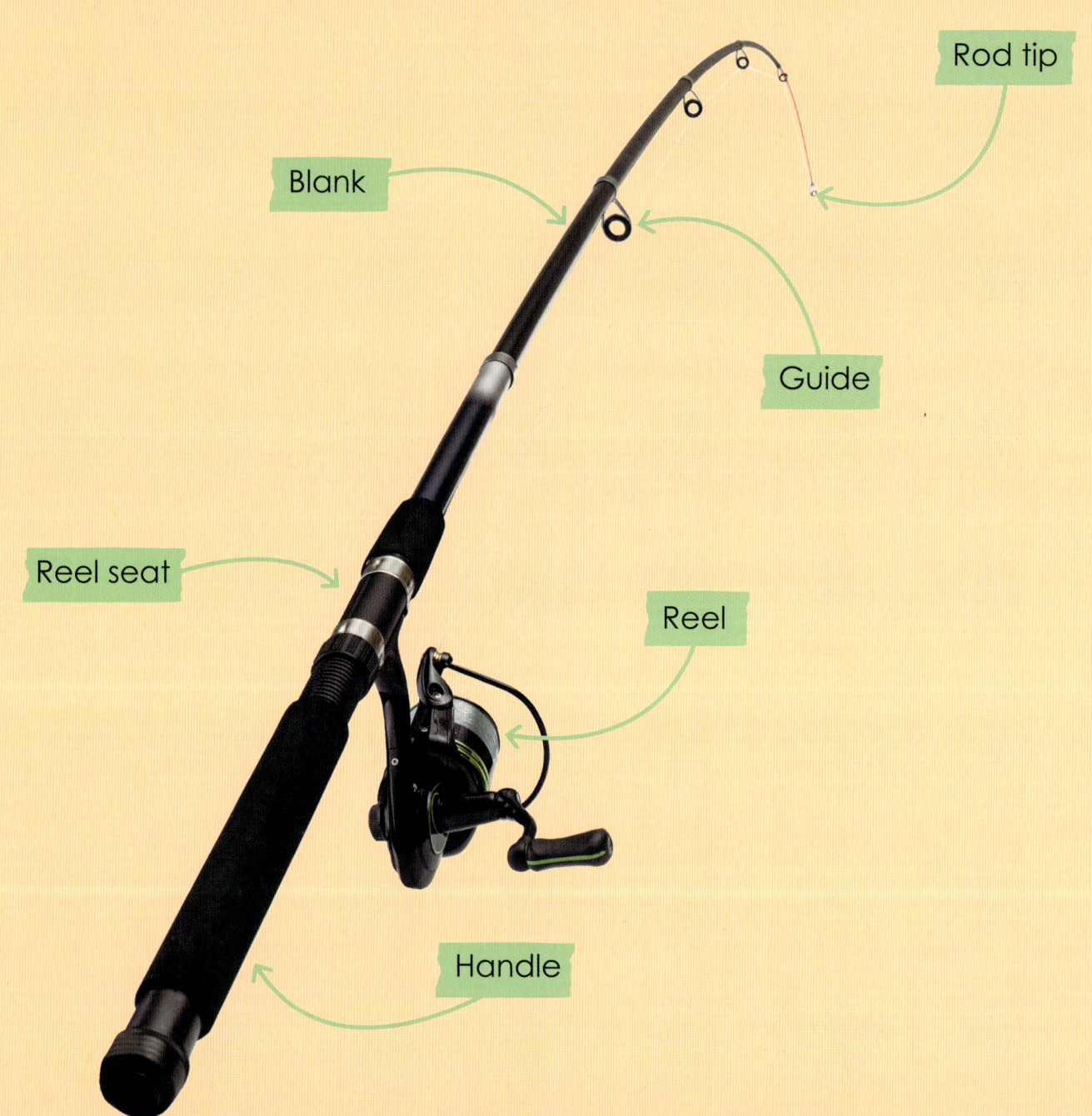

Rod tip

Blank

Guide

Reel seat

Reel

Handle

The speed of the reeling depends on the kind of fish the angler is trying to catch. Some species prefer to chase a fast lure, while others like a slower one.

The speed also affects how deep the lure will be. Faster reeling keeps it shallower while slower reeling keeps it deeper.

IF YOU DIDN'T GET A BITE, JUST CAST AGAIN!

A DAY AT THE LAKE

**It's a great day to be out in nature!
Why not fish for some carp at the lake?**

Look for any active areas of the lake. You may even see some carp jumping out. Unpack your rods nearby and set up your equipment.

On each rod, you can run the line through an electronic **bite** alarm. When a fish pulls the line, it sets off the alarm.

Bite alarm

Use bait such as **feed pellets** or **boilies** inside a special bag. The bag will **dissolve** and leave the pellets on the lake bottom. They will attract the carp to the bait.

Fishing takes patience. But the excitement starts as the alarm to reel the fish in goes off!

FLOATS AND SINKERS

When spin fishing, there are two main methods that anglers use – floats and sinkers.

Floats are used to present a bait between the lake bed and the surface.

They are good for catching **predatory** fish such as pike, zander, perch and catfish. These fish like shelter, so they hide in snag trees and reeds.

Sinkers, also known as legers, are weights that hold the bait in one place on the bottom of the lake. Sinkers are usually made from lead, brass or steel.

Sinkers help anglers reach fish that stay near the bottom of the lake bed, such as bream.

A sinker attached to a fishing line

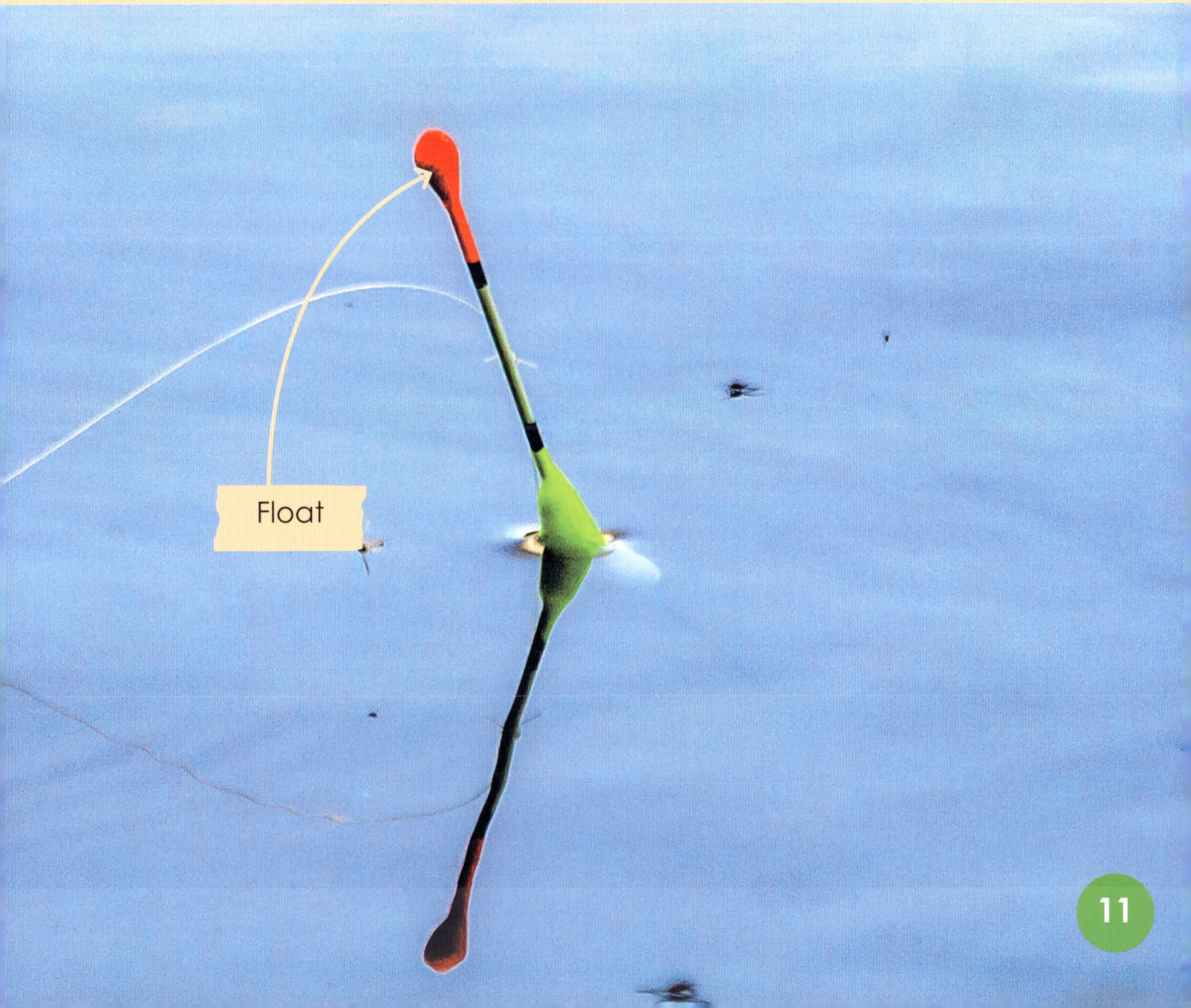

Float

BAIT AND LURES

The choice between using bait or lures depends on the fish species and its feeding habits. Sometimes anglers use both.

LOTS OF DIFFERENT BAITS CAN BE USED FOR SPIN FISHING:
- Worms and maggots
- Bread, cheese and sweetcorn
- Specially made boilies

Boilies for carp fishing

Lures are made to look like real fish. It's best not to use plastic lures. Instead, you can get lures that break down naturally in the water.

Lures

Different lures attract different fish. Always carry a selection so you can try different ones.

Spoons reflect the light as they wobble through the water.

Silicon lures are flexible. They move and flow naturally in the water.

Spinners spin around a fixed stem. This movement attracts the fish.

CARING FOR THE CATCH

You can return fish to the water after catching them. This is known as catch-and-release fishing. It preserves the supply of fish.

Always use a landing net to **land** the fish.

Keep the fish out of the water just long enough to unhook, weigh and photograph the fish.

It is important that fish go back into the water in the same condition as when you caught them.

Unhooking mat

Big fish should be laid on a padded unhooking mat while you gently unhook them.

If you don't want to release the fish straight away, you can keep it in a floating basket or **keepnet** for a short time.

Fish-friendly basket to keep live fish

FLY FISHING

Fly fishing is one way to catch fish such as trout and salmon. Anglers use a long, responsive rod to move the fly around.

In clear water, the angler can spot the fish easily and cast the fly directly to it.

Fly fishing gets its name from the lures that are used. Anglers use small lures that look like insects that fish may want to eat.

The technique of fly fishing is thought to have originated almost 2,000 years ago!

FLY FISHING – CASTING

To make a basic fly fishing cast, work the rod backwards and forwards to get the right amount of line.

First, grip the rod as if you are shaking someone's hand. Pull as much line as you think you need off the reel.
This will sit in a pile at your feet.

Hold the rod in front of you. Then pull it back until the line straightens behind you.

Now quickly bring the rod forwards and wait for the line to straighten out in front of you.

Taking the line

Keep moving the rod back and forth to allow more and more line out through your fingers.

When your line is long enough to reach your target, let the fly drop onto the water.

FLIES AND GEAR

Flies are made from fur, feathers and thread tied onto a sharp hook. They copy the look and movement of a real insect. There are 3 fly fishing types:

NYMPH
- Used below the water's surface
- Great for beginners

DRY FLY
- Used on the water's surface
- Best in shallower water

STREAMER
- Used deeper below the water's surface
- Takes more skill but catches bigger fish

Whip finish tool

Tying flies is very fiddly! But it's great to catch a fish with a fly you've tied yourself.

Fly fishing anglers wear vests with lots of pockets for storing gear such as fly boxes and sunglasses.

They also wear long, rubber boots or all-in-one boots and trousers. These are known as waders.

Vest

Waders

SEA FISHING

Some sea anglers like to fish for large fish, such as marlin, tuna and shark. This type of sea fishing is known as big game fishing.

Black marlin can swim up to 80 mph!

Marlin are large, powerful fish. This makes them difficult to catch.

The largest ever marlin caught on a rod and reel weighed 707 kg and was 4.4 metres long!

Black marlin

CATCHING A MARLIN

Big game fishing is done from a boat. Anglers can charter a boat with an experienced crew.

Marlin anglers use extra strong rods with thick line. They use lures that look like squid – the marlin's main food.

Anglers can also use natural bait, such as mackerel, combined with a lure. These are called "skirted baits".

Reels

Big game rods

Marlin can be found in warm oceans in places such as the Caribbean, South America, Africa, Australia and New Zealand.

A New Zealand striped marlin

A string of **teasers** is pulled behind the boat. This is known as trolling. The main lure is on the end, so it is first to be hit by the marlin.

ICE FISHING

Cold weather doesn't stop anglers fishing. They just drill into the ice and keep enjoying their hobby!

To get to the fish, anglers make a hole in the ice with an ice auger.

For safety, the ice must be a minimum of 10 cm thick. If snowmobiles or other vehicles are on the ice, it should be much thicker.

Hand ice auger

Power ice auger

Next, they put bait or lures at the end of light fishing rods and drop them into the water.

Anglers can have many baited lines at once by using tip-ups. Tip-ups are attached to the line. When a fish takes the bait, it pulls the line and makes the flag go up.

The tip-up is set.

A fish has been caught!

COMPETITIVE FISHING

Anglers can show off their skills in fishing tournaments around the world.

Different tournaments have different rules, such as counting:

- the total number of fish caught

- the total weight of fish caught

- the total weight or length of an angler's top five fish

Some tournaments give out enormous prizes. In 2023, the White Marlin Open (Ocean City, Maryland, USA) gave out a top prize of nearly £5 million!

The fishing pier in Ocean City, Maryland

CATCHING A RECORD

There are two types of records for anglers – national records and IGFA records.

- If you catch a record-breaking fish, you will need someone to witness your catch. Another angler is best.
- Get your scales checked to make sure they are accurate.
- Weigh the fish in front of your witness. Contact the record-keeping organisation to let them know.
- Fill out the claim form!

IGFA stands for "International Game Fish Association".

GLOSSARY

angler – a person who catches fish using a rod and line.

anticipation – the feeling of looking forward to something.

bait – food used to attract fish.

bite – when a fish takes the bait.

boilies – specially made fishing bait. Boilies are balls of paste with different flavours. They can be bought from fishing shops.

cast – to throw one end of a fishing line into the water.

charter – to hire a boat and its crew.

dissolve – to mix with water or another liquid and become part of it.

feed pellets – small, hard balls of food.

imitation – something that looks like, or pretends to be, something else.

keepnet – a net for keeping fish alive in the water until they are let go.

land – to catch a fish and bring it out of the water.

predatory – the word to describe an animal that hunts and eats other animals.

reel in – to pull a fish out of the water by winding up the line.

responsive – something that responds well. A responsive rod will make the actions you need quickly and accurately.

teaser – a lure or piece of bait that is used to attract a fish, but that doesn't have a hook.

INDEX

B
bait 5, 9, 10-11, 12, 24, 27, 31
big game fish 5, 22, 24-25
big game fishing 22, 24-25
bite alarm 8-9
boats 5, 24-25, 31
boilie 9, 12, 31
bream 11

C
caring for a fish 14-15
carp 8-9, 12
casting 6-7, 16-17, 18-19, 31
catch-and-release fishing 14-15
catfish 10
competitions 28-29

D
dolphinfish 5

F
flies 5, 16-17, 20-21
floats 10-11
fly fishing 5, 16-17, 18-19, 20-21

I
ice augers 26
ice fishing 26-27
IGFA (International Game Fish Association) 30

L
legers 10-11
line (fishing) 6, 8, 11, 18-19, 24, 27, 31
lures 6-7, 12-13, 24-25, 27, 31

M
mackerel 24
mahi-mahi 5
marlin 22-23, 24-25, 29

P
perch 10
pike 10

R
record-breaking fish 30
reels 4, 6-7, 18, 22, 24, 31

rods 4, 6, 8, 16, 18-19, 22, 24, 27, 3

S
salmon 16
sea fishing 5, 22-23, 24-25, 28-29
sharks 22
sinkers 10-11
spin fishing 4, 6-7, 10-11, 12-13

T
teasers 25, 31
tip-ups 27
trolling 25
trout 16
tuna 22

W
waders 21

Z
zander 10

Picture credits:
(t=top; b=bottom; m=middle; l=left; r=right):
Shutterstock: WBMUL 1bg; Natalia Kirichenko 4bg; Krasula 5t; Somprasong Khrueaphan 5b; Gresei 6bg; FtLaud 7b; Alexey Masiliy 11tr; Buhta Yurii 12tr; OlegDoroshin 12b; Mriserg 13t; Nikamo 13br; Maksim Lysyuk 13bl; FedBul 14b; MatrukIstvan 15t; Gennady Grechishkin 15br; Goodluz 16-17bg, 31b; Warren Claflin 18m, 19m; Anders Worre 19tr; Allison Achauer 21br; Alejandro Vicente 21t; Joshua Rainer Photography 20tr; NPF Photography 20br; Tammy Goodwater 20ml; Kelldallfall 22-23bg; Swordfishmike 25tr; Go2dim 25b; Corywheeler 24b; Project1photography 24tr; Stephen Mcsweeny 26mr; Maxim Petrichuk 26ml; Yermolvych 27mr, 27ml; Fabien Monteil 28-29bg; Jon Bilous 29tr; irinaK 30tl; Felix Mioznikov 30b; AleMasche72 8t, 8b, 9bg; Cindy Creighton 2-3bg; Akiyoko 10-11bg.

Every effort has been made to trace the copyright holders, and we apologise in advance for any unintentional omissions. We would be pleased to insert the appropriate acknowledgements in any subsequent edition of this publication.